W9-ADE-867

Understanding the Elements of the Periodic Table™

THE TRANSITION ELEMENTS

The 37 Transition Metals

Mary-Lane Kamberg

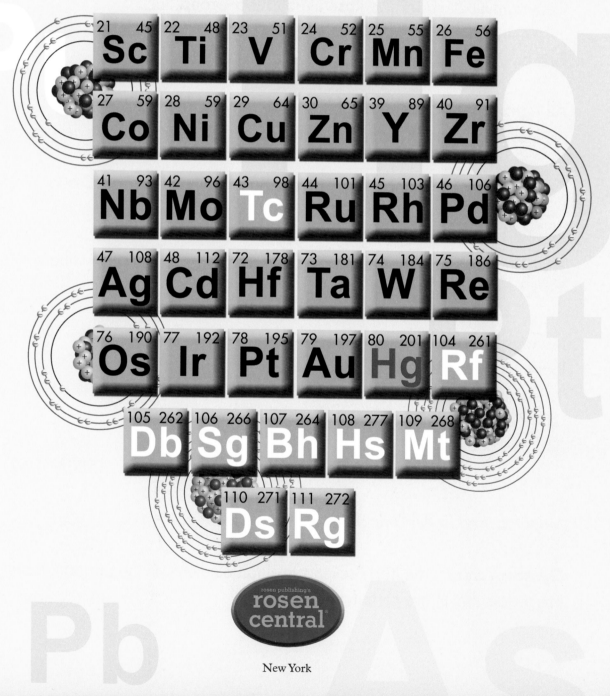

rosen publishing's
rosen central®

New York

Dedicated to the teachers in my family: Franck Brichet, Cindy Clenard, Eric Falls, Ruth Ann Falls, Ricki Gilbert, Melinda Oakes, Susan Shanaman, and Brynn White
And in memory of Lena Kamberg and Bobbie Griffith Powell

Published in 2010 by The Rosen Publishing Group, Inc.
29 East 21st Street, New York, NY 10010

Library of Congress Cataloging-in-Publication Data

Kamberg, Mary-Lane, 1948–
The transition elements: the 37 transition metals / Mary-Lane Kamberg.—1st ed.
    p. cm.—(Understanding the elements of the periodic table)
Includes bibliographical references and index.
ISBN-13: 978-1-4358-5332-4 (library binding)
1. Transition metals—Popular works. 2. Periodic law—Popular works. I. Title.
QD172.T6K36 2010
546'.6—dc22

2009001052

Manufactured in the United States of America

On the cover: The thirty-seven transition elements of the periodic table are metals that share certain characteristics.

Contents

Introduction

Have you ever gazed into the night sky and wondered what the universe is made of? Not long ago, legend said the moon was made of green cheese.

Today, however, scientists know the universe is made of energy and matter. Energy is the ability to do work. Energy may create heat, electricity, and light. Matter is "stuff" that has mass and takes up space. Matter can be a solid, liquid, or gas. This "stuff" is made of atoms.

An element is a substance that's made up of only one kind of atom. Only ninety-four elements are found in nature on Earth. But at least 115 elements are known, and more are predicted. The extra ones have been produced by scientists in experiments. Together, the natural elements and the ones produced in laboratories make up the matter of the universe.

Most elements are metals. The rest of the elements are either non-metals or metalloids. Chemists arrange the elements in a chart called the periodic table. The rows of the periodic table are called periods. Elements in the same horizontal row have little in common.

However, the elements also line up vertically. The vertical columns are called groups (or families). Elements lined up in the same group have similar chemical and physical characteristics. Their atomic structures (see chapter 2) determine these characteristics. If you know which group an

Transition elements often form colorful compounds. From left, the flasks above contain compounds of titanium, vanadium, chromium, manganese, iron, cobalt, nickel, and copper.

element belongs to, you can make an educated guess about what the element is like—even if you have never heard of it or seen it.

Elements in the first vertical column or group—except hydrogen, which is a nonmetal—are called alkali metals. Group 2 elements are called alkaline earth metals.

Groups 13–16 contain a combination of nonmetals, metalloids, and other metals.

Group 17 represents the halogens. Group 18 elements are called noble gases. They seldom combine with other elements.

The two horizontal rows at the bottom of the table are known as inner transition metals. They include the lanthanide series and the actinide series. Together, the lanthanides and actinides are known as rare earth metals. They were once thought to be rare on Earth. Actually, they were only difficult to separate into pure form. Even though separating them is easier today, the name is still used. The metals in these two rows share so many characteristics that they're hard to tell apart. Except the actinides have an extra characteristic: they are radioactive. They're used in nuclear fuel, smoke detectors, and atom bombs.

That leaves groups 3–12. They are transition elements, also called transition metals. These metals have their own rules. You can't even count on some of them to act the same way every time they combine with other elements. And, from one element to the next, they aren't much different. They all share similar properties and behave in similar ways.

Most of the metals you encounter every day are transition metals. Let's learn more about them.

Chapter One
How It All Works

Have you ever looked in a mirror? Used a spoon to eat raisin nut bran? Or dropped coins into a vending machine? If so, you have used transition metals.

Mirrors may be coated with silver (Ag) or rhodium (Rh). Stainless-steel "silverware" is made of iron (Fe) and chromium (Cr). Raisin nut bran contains manganese (Mn), iron, copper (Cu), and zinc (Zn). Coins are made of gold (Au), silver, nickel (Ni), copper, and zinc.

Transition metals are shiny and strong. They conduct heat and electricity. You can pound them into thin sheets or draw them out into wire.

Some are common. You've certainly heard of silver, gold,

Many countries, including the United States and Canada, use transition elements like gold, silver, nickel, copper, and zinc to make coins.

and copper. But do you know about darmstadtium (Ds), roentgenium (Rg), or osmium (Os)? Some transition metals are among the earliest elements known. More than ten thousand years ago, people knew about copper. They used it for tools, weapons, and jewelry. Gold ornaments dating from 6000 to 4000 BCE have also been found.

From the Stone Age to the Iron Age

The first human cultures used stone for tools and weapons. This period is known as the Stone Age. At the end of the Stone Age, humans began using the transition element copper for these purposes.

Around 3000 BCE, metallurgists learned that heating metals makes them easy to mix together. They combined copper and tin (Sn) to make bronze. Bronze made stronger tools and weapons than copper alone. Soon bronze swords, knives, hammers, and arrowheads appeared. The time between 3000 BCE and 1200 BCE is known as the Bronze Age.

By the end of the second millennium BCE, people developed techniques to purify iron from its ore. About 1200 BCE, the Iron Age began. Pure iron is a strong but soft element. Metallurgists experimented with combinations of iron and different amounts of carbon (C) to strengthen the material. Adding carbon to iron makes steel, which is much stronger than iron. Steel made tools and weapons that were stronger than those made from bronze.

Early Transition Metals

The first known transition metals include gold, silver, and copper. These three elements occur in pure form in nature, so they were easy to find and use. (Most of the other transition metals exist only combined with other elements or substances. They have to be separated before they can be used.) Ancient people also discovered iron and mercury (Hg).

Ancient Egyptians replaced weapons and tools made of copper with ones made of bronze. Adding tin made stronger objects than those made of copper alone. This innovation brought on the Bronze Age in Egypt during the second millennium BCE.

Fire played a key role in discovering new transition metals. That's because most metals are found in ores. Ores are minerals or groups of minerals that contain concentrated deposits of desired elements, particularly metals. Fire heated ore. The heat helped separate the desired metal from the other substances. Fire also let metallurgists mix metals together. Metallurgists work to purify metals and find uses for them.

Finding New Elements

As ancient metallurgists worked, they came upon new substances. But they didn't recognize them as elements. As recently as 100 BCE, ancient Egyptians thought only four elements existed: earth, air, fire, and water. They weren't looking for any more elements. But they were looking for more of *something*: gold.

Through a practice called alchemy, they tried to make gold from other metals. Alchemy spread from Egypt to Greece, Syria, and Persia (now Iran). It became associated with magic and spirits. No one ever succeeded in making gold. But as they tried, they created a body of knowledge about mixing different metals. They also developed equipment

How Transition Metals Are Alike

Transition metals share many properties. They are:
- Shiny
- Dense
- Good conductors of heat and electricity
- Able to be pounded into thin sheets or drawn into wire
- Good catalysts (they help start or speed up chemical reactions without being consumed)

and methods still used in chemistry laboratories today. And even though alchemists weren't looking for new elements, their work helped later chemists discover some. The most important was oxygen (O).

Oxygen Leads to More Discovery

Before the discovery of oxygen in 1772, the only known transition elements were gold, iron, mercury, silver, platinum (Pt), and zinc. Most of the rest of the transition metals were locked in combinations with other elements, especially oxygen. However, between 1781 and 1808, scientists used chemical reactions to separate oxygen from other materials.

Scientists isolated manganese in 1774, just before the dawn of the American Revolution. The discoveries of molybdenum (Mo) in 1782, tungsten (W) in 1783, zirconium (Zr) in 1789, titanium (Ti) in 1791, yttrium (Y) in 1794, and chromium in 1797 soon followed.

In 1801, the same year U.S. president Thomas Jefferson took office, scientists discovered niobium (Nb) and vanadium (V). Tantalum (Ta) came along in 1802, followed by iridium (Ir), palladium (Pd), and rhodium in 1803 and osmium in 1804. Scientists were also discovering new nonmetallic elements in other groups. They needed a way to sort them all out.

The Periodic Table

In 1869, a Russian chemistry professor named Dmitry Mendeleyev published his first table putting the known elements in order. He and others later perfected his invention as the periodic table of the elements. Mendeleyev arranged the known elements by their atomic weights. He noticed that every eight elements' characteristics repeated. So he made columns of elements that shared the same properties.

Dmitry Mendeleyev created the first version of the periodic table of the elements. Since publication of his *Principles of Chemistry*, other scientists have added to and changed the table, which is still used today.

As additional elements were discovered, they had a proper place on the periodic table. The new elements included scandium (Sc), discovered in 1879; hafnium (Hf), discovered in 1923; rhenium (Re), discovered in 1925; technetium (Tc), discovered in 1937; rutherfordium (Rf), discovered in 1964; dubnium (Db), discovered in 1967; and seaborgium (Sg), discovered in 1974.

The Father of the Periodic Table

Although other scientists were working on a table of the elements at the same time, Dmitry Mendeleyev, a scientist and professor born in Siberia in 1834, gets most of the credit for its invention. He arranged the elements from lightest to heaviest. Only sixty-three were known at the time, but he left spaces where additional elements should have been. He published his table in *The Principles of Chemistry* in 1869. Later scientists modified the table based on new discoveries.

Producing New Elements

Between 1981 and 1994, scientists at the Gesellschaft für Schwerionenforschung (GSI) laboratory in Darmstadt, Germany produced five new transition elements. Elements discovered at the research center include bohrium (Bh) in 1981, meitnerium (Mt) in 1982, hassium (Hs) in 1984, darmstadtium in 1994, and roentgenium in 1994. (Some scientists consider ununbium, discovered in 1996, to be a sixth element produced at this laboratory.) The facility has a particle accelerator that uses electrical fields to increase the energy of atomic particles so nuclear reactions that produce new elements can occur.

Chapter Two
Break It Down

Why are transition metals classified together? The answer depends on the elements' atomic structures. An atom is the smallest unit that has the characteristics of an element.

The basic parts of an atom are called subatomic particles. The most important subatomic particles are neutrons, protons, and electrons. The center of an atom—called a nucleus—contains protons and neutrons. Electrons spin around outside the nucleus.

All neutrons are the same. All protons are the same. And all electrons are the same. It is the number of subatomic particles and the energy levels of the electrons and their placement in an atom that create the element's properties.

This illustration of atomic structure shows electrons fairly close to the nucleus, which contains protons and neutrons. However, the distance between electrons and the nucleus is much greater. In fact, much of an atom's structure contains empty space.

It's a Small World

Atoms are quite small. You could fit one hundred million hydrogen (H) atoms across the width of your fingernail.

Compared to atoms, the subatomic particles they're made of are even smaller. Protons are about the same mass as neutrons. Electrons are much smaller—about 1/2000 the size. (So it would take about two thousand electrons to equal the mass of a proton or neutron.)

In addition to being small, most of the mass of the atom is squeezed into a tiny space, the nucleus of the atom. If the nucleus of a hydrogen atom were the size of a pea, the space occupied by the entire atom would be as big as the Superdome in New Orleans. The pea (the nucleus)

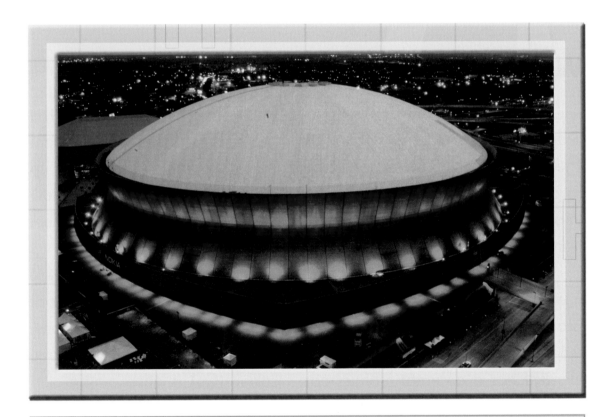

If the Superdome in New Orleans, Louisiana, were the size of one hydrogen atom, its nucleus would be the size of a pea. The dome itself has a diameter of 680 feet (210 meters).

15

would be suspended in the middle of the stadium, which seats almost 73,000 football fans! The hydrogen atom's single electron would dart around filling all the rest of the space inside the stadium.

Electrons and protons have electrical charges. Electrons have a negative charge. Protons have a positive charge. Particles with the same charge push each other away. So two electrons (both negative) move away from each other. However, particles with different charges attract each other. So one electron and one proton come toward each other.

The attraction helps explain why an atom holds together. Even though an electron is far away from a proton, the electron's negative charge attracts the proton's positive charge, holding the electron in the atom. Neutrons have no electrical charge. They're neutral. So is the complete atom the subatomic particles belong to.

Weighing In

An element's atomic number is the number of protons in its nucleus. (The order of elements in the periodic table is according to the atomic number.) The total mass of the protons, neutrons, and electrons in an atom make up its atomic weight (also known as atomic mass). Now you can see how important subatomic particles are. By adding one proton (and an electron) to an element, you get an entirely different, heavier element.

In some cases, the same element can have atoms with different numbers of neutrons. An isotope is an atom that has the same number of protons as other atoms of the element, but a different number of neutrons. Adding a neutron to an atom adds weight. So an isotope of the same element has a different atomic mass. Cobalt (Co) is an example of a transition element with isotopes.

No matter where they are, the same form of an element's atom is the same whether it's on Earth, Mars, the moon, or in deep space. An atom of each particular isotope of an element is also the same wherever it's found.

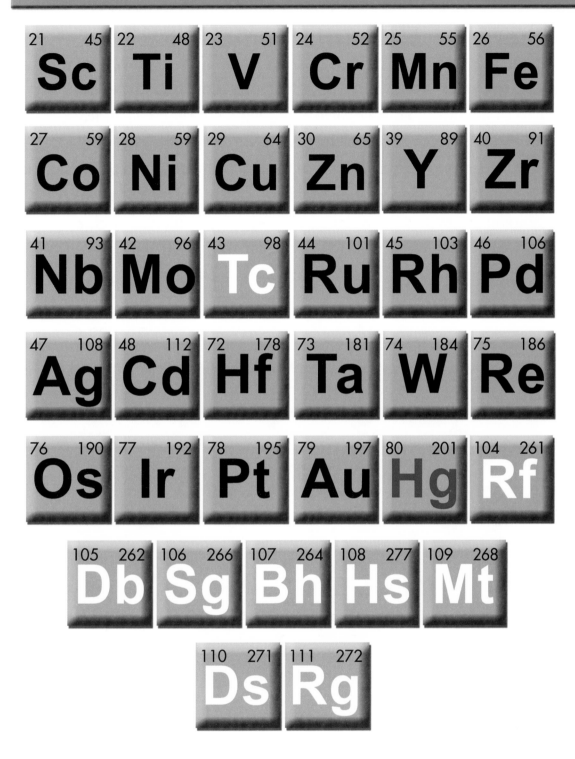

The number in the top right corner of each element on the periodic table represents the atom's atomic weight, which is the total mass of the protons, neutrons, and electrons in an atom.

17

Where Do Elements Come From?

In ancient times, some groups of humans worshipped the sun as the source of the universe. As it turns out, they were on the right track.

The sun is a star. Many elements are formed in the cores of stars.

Hydrogen and helium (He) were the first elements in the universe. They still amount to 97 percent of all matter. A star burns hydrogen into helium by joining together the nuclei of two hydrogen atoms. This process is known as fusion. When all the hydrogen burns off, helium burns into oxygen, then oxygen burns into carbon and oxygen, and so on. As the

Stars usually don't burn hot enough to create elements heavier than iron. But when a dying star explodes as a supernova, nuclear reactions generate enough heat to form new elements. The resulting matter is cast into space in an expanding debris field.

atoms combine, they create heavier and heavier elements. Over the life of the star, elements up to and including iron form.

Normally, a star can't burn hot enough to create the nuclear reactions that create elements heavier than iron. But when an old star "dies," it explodes in a spectacular burst of light and energy known as a supernova. A supernova creates the conditions needed to make elements heavier than iron. The new elements then fly off into space.

Scientists in laboratories like the GSI research center in Germany use experimental nuclear reactions to produce new transition elements. They make only small quantities that last only a matter of seconds. Scientists have little chance to observe their properties.

Atomic Structure

The arrangement of an atom's subatomic particles in an atom is its atomic structure. Atomic structure is like a mini solar system. However, electrons don't orbit the nucleus the same way planets orbit the sun. Instead, they move around like a swarm of gnats. Still, there is order in the madness.

Electrons have different energy levels at different distances from the nucleus. Those closest to the nucleus have the lowest energy level. Those

More Ways Transition Metals Are Alike

Transition elements have high melting and boiling points. All except mercury are solid at room temperature. When a new transition element forms, the added electron goes to the second-to-last subshell, not the last. Transition elements have valence electrons in the last two subshells.

farthest away have the highest. Electrons also spin in different directions, and the shapes of their orbits differ. Scientists don't know exactly where electrons are at a particular time, but they do know where they are likely to be. The term "electron cloud" refers to these places.

The energy levels where electrons are found are known as subshells. Each subshell can hold a specific number of electrons. For example, one type of subshell can hold only two electrons. Another type holds up to six.

How Transition Elements' Structures Are Different

When most new elements are formed, a proton joins the nucleus of an atom. The corresponding electron goes to the first atom's outermost subshell, farthest from the nucleus. Formation of a transition element breaks the pattern. The outermost subshells of transition elements are full. But the second-to-last one has room for more. New electrons join not the last subshell, but the second-to-last one.

Chapter Three
Finding and Using Transition Elements

Many transition elements are found in nature in concentrated ore deposits. An ore is a combination of minerals that contains a commercially valuable metal. So while prospectors called forty-niners during the California gold rush of 1849 found loose gold nuggets in streambeds, gold and other elements also exist in ore deposits or bound together with other elements.

Refining Transition Elements

Most transition elements have uses in pure form. Before the metals can be used, however, they must

During the California gold rush, prospectors used a method known as cradling to separate gold from ore by washing. Washing is one of many methods that is used to separate transition elements from the ore or compounds where they are found.

be separated from the ores or compounds where they are found. Workers must refine the element by reducing it to its pure form.

Smelting

Perhaps the earliest known way to separate transition metals from ore dates to at least 3500 BCE. Using fire, a process called smelting separates pure metal from the rest of an ore by making reduction reactions possible.

For example, heating carbon with an ore that contains a compound of iron and oxygen starts a series of chemical reactions that combine the oxygen in the compound with the carbon. The result is carbon dioxide and iron. The reaction is called a reduction reaction because the original ore is reduced to the pure element. The pure metal weighs less than the original ore. The mass is reduced in the process of purifying the ore.

If other impurities remain, metal workers make a foamy substance called a flux, which melts together with the rest of the impurities. The impurities are less dense than the iron. Therefore, the impurities float. Workers then skim off the impurities, leaving the pure element. Smelting is used to purify copper, iron, and other metals.

Dangerous Heavy Metals

Chromium, cadmium (Cd), and mercury are "heavy metals," which are dense, poisonous elements. Heavy metals are not eliminated from the body after digestion. If a fish eats something containing mercury, the mercury stays in its body. If a bigger fish eats the first fish, the poison stays in the bigger fish—and in a bear or human that eats it. This process is called bioaccumulation (bio for life; accumulation for collecting the poison), a dangerous result of pollution.

Distillation

Ancient alchemists developed another way to separate metals from impurities. One way that separates substances in a mixture is called distillation. This process takes advantage of the different boiling points for different substances. The scientist heats the mixture until one of the substances in it evaporates, and the vapor is condensed into a pure substance. (The evaporated material can also be cooled back to a liquid, if desired.) Tungsten is often purified using a special form of this method.

The apparatus used in the process of distillation originated in ancient times when alchemists tried (but failed) to turn other metals into gold. Today, the method helps get rid of impurities and separate elements in a mixture.

Amalgamation

Another method from ancient times is amalgamation, which uses mercury to refine gold and silver. During the California gold rush, miners added mercury to crushed ore. The mercury blended with the gold in the crushed ore to form an amalgam. An amalgam is simply a mixture of mercury and another metal. The amalgam was easily separated from other substances in the ore. Miners then got rid of the mercury in the amalgam by heating it. Mercury vaporizes at a lower temperature than gold. Pure gold was left.

The trouble with this method is mercury is highly poisonous. Mercury intoxication was a common ailment among gold rush prospectors in the United States and Canada. Some people died from breathing mercury's vapors. Mercury contamination from mining in the late nineteenth century still pollutes the environment today. So amalgamation to recover gold is limited to small operations in the United States. However, amalgamation is still used in some developing countries.

Washing

Another way to get a transition metal out of an ore that contains pure metal is to wash it. First, miners crush the ore and pour it onto a shaking, sloping table as water washes over the material, separating the metal from the other substances in the ore. During the gold rush era, many prospectors used this method in search of gold.

Froth Flotation

Froth flotation is another refining process. Workers add water to crushed ore and blow air through the mixture. They add a substance that helps the unwanted materials attach to bubbles of air. As the air bubbles rise to the surface, they take the ore particles with them. Workers skim off the material, leaving the metal at the bottom. Zinc is one transition element that is purified this way.

Electrolysis

In a molten or dissolved substance that contains both metallic and non-metallic particles, a process called electrolysis separates the particles to isolate a desired transition metal like copper or niobium. This method uses electric current to create chemical reactions that isolate the metal.

Magnets

Magnets can separate some metal ores from crushed rock. Magnetism is a force that attracts or pushes away objects at a distance. Magnets attract

A toy magnet can pick up objects made of iron, nickel, or cobalt. Large industrial magnets are used in mining to separate metal from ore. The recycling industry also uses large magnets to separate metal cans.

such metals as iron. If a magnet is passed over a quantity of crushed rock, the magnet attracts the metal, leaving other substances in the rock alone.

Uses of Transition Elements

Once a transition element has been purified, it's ready to use. And transition elements serve a wide variety of purposes in manufacturing, health care, and everyday life. In industry, many transition metals serve as catalysts for chemical reactions. A catalyst is a substance that helps start or speed up chemical reactions.

Platinum and vanadium, for example, are catalysts used in reactions that produce sulfuric acid. Rhodium is used in catalytic converters to clean automobile exhaust. And palladium helps purify hydrogen.

"Attractive" Toys and Machine Parts

Iron, nickel, and cobalt are three magnetic transition metals. That means they can attract iron and other metals. Magnets make fun toys. They're also used in such products as headphones, computer speakers, credit card magnetic strips, garbage disposal motors, doorbells, refrigerator ice makers, garage door openers, electric toothbrushes, and more.

Transition Metals in Health Care

In the health care industry, titanium is used in artificial hips and heart pacemakers. Technetium helps doctors find tumors, and a form of cobalt helps them treat cancer. So does iridium, which helps control tumors. Platinum is used in anticancer drugs. Another transition metal called tantalum is used

This X-ray shows an artificial shoulder joint made of the transition element titanium. Titanium also makes fireworks silver. And a compound of titanium and oxygen called titanium oxide makes paint white.

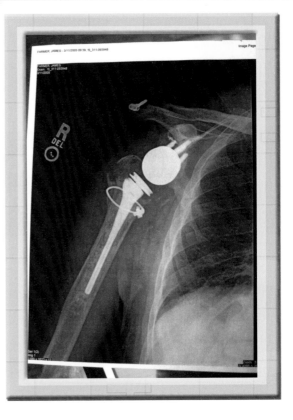

in surgical appliances. And some thermometers contain mercury. Dentists sometimes use gold to make fillings.

Some transition metals are important for good health. Chromium helps regulate the amount of sugar in blood, which is important to people with diabetes. The body uses iron to produce hemoglobin, the molecule that transports oxygen in blood. (However, iron can be dangerous. Too little causes anemia, which makes a person weak. Too much damages the kidneys.)

So far, no uses are known for the transition elements produced in nuclear experiments. However, they may have future use in medical treatments.

Chapter Four
Mixing It All Up

When two or more elements are involved in a chemical reaction, the electrons involved in the chemical reaction are called valence electrons. Valence electrons have high energy levels, and their attraction to the protons in the atom's nucleus is weak. So it's easy for them to be attracted to the nucleus in another atom. In a chemical reaction, the elements give up, gain, or share valence electrons. In most elements, valence electrons are found in the last subshell—farthest from the nucleus. In transition elements, however, valence electrons come from the last two subshells.

The result of a chemical reaction is an entirely new substance called a compound. The original elements cannot be separated again without another chemical reaction. The subatomic particles of the original elements are still there, but they have a new arrangement.

Metal atoms, including those of the transition elements, tend to lose electrons. Nonmetals tend to gain them. When transition metals react with nonmetals, the new substance is a salt. Salts have the structure of a crystal, a 3-D form made of identical, repeating building blocks.

The table salt you shake onto French fries is a combination of sodium (Na), which is an alkali metal, and chlorine (Cl), which is a halogen and a nonmetal. That is one type of salt. But you won't find salts made from chemical reactions with transition metals on your kitchen table! They have other uses.

Halite, more commonly known as rock salt, is a compound of sodium and chlorine. Notice the repeating building blocks that create the crystalline structure characteristic of salts. Salts are also formed by combinations of transition elements and nonmetals.

Silver compounds made with bromine (Br), chlorine, and iodine (I) are used in photography and motion pictures. Silver iodide is used to "seed" clouds to produce rain. Tantalum oxide is used to make high-quality camera lenses. Tungsten salts are used in tanning leather. When salts made from transition metals are melted or dissolved in water, they conduct electricity, which offers many uses in manufacturing and other industries.

In health care, a compound of osmium and oxygen treats arthritis. Another compound made from iron, phosphorus (P), and other elements strengthens red blood cells and controls bleeding.

Precious Gems

Gemstones are admired for their beauty and used for jewelry. Gemstones are minerals in crystal form. Some gems get their color from transition elements. In one compound, chromium creates the red color of a ruby; in another, it makes an emerald green. Titanium gives a sapphire its shades of blue. Iron makes topaz yellow.

How Do Compounds Form?

Reactions of metals and nonmetals create ionic bonds. The bonds hold the subatomic particles of the elements together as a new substance. In forming an ionic bond, a metal atom gives up one or more electrons. The electrons join the outermost energy level of a nonmetal atom. The protons that were paired with the lost electrons stay with the metal atom. The metal atom is now called an ion. An ion is an atom with unequal numbers of protons and electrons.

An ion is no longer neutral. Remember, an atom has the same number of positively charged protons and negatively charged electrons. In the ion, though, some of the negatively charged electrons are gone. With more protons than electrons, the metal ion now has a positive charge.

The opposite happens with the nonmetal atom that gained the electrons the metal atom gave up. The nonmetal atom is also an ion because the number of electrons is greater than the number of protons in the atom. The nonmetal ion got more electrons but no protons. Electrons have a negative charge. There are more electrons than protons, so the nonmetal ion has a negative charge.

Here's where the "opposites attract" rule comes into play. The positively charged metal ion is attracted to the negatively charged nonmetal ion. The two join in a new arrangement of subatomic particles. The two types of ions do not "pair up." Instead, each metal ion has the same attraction for all the nonmetal ions, and vice versa. The result is a neutral substance (since the whole now has an equal number of protons and electrons). Like an atom, the compound that results from the chemical reaction must be neutral—with equal numbers of positive and negative charges.

Most metal atoms lose the same number of electrons, no matter which nonmetal they react with. But transition metal atoms have their own rules. Transition metals can lose different numbers of electrons in different situations. For example, iron can lose either two or three electrons in chemical reactions.

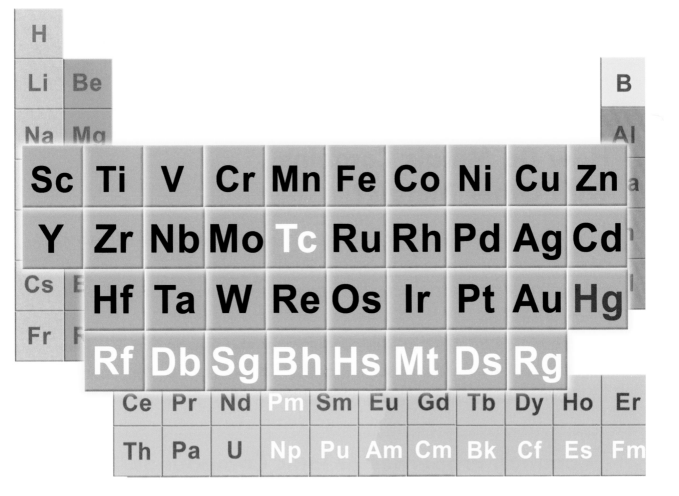

When metals combine with nonmetals, the resulting compound is a salt. Metals do not form compounds with other metals, but they can be mixed together to form alloys.

This ability helps transition metals easily bond with other elements. Copper reacts with oxygen to form copper oxide and copper dioxide. It also reacts with carbon, chlorine, hydrogen, and sulfur (S) to create copper carbonate, copper chloride, copper hydroxide, and copper sulfate.

Transition elements often make colored compounds with a variety of uses. Yttrium oxide makes the red in picture tubes for color televisions and computer monitors. Metal salts made from chromium, iron, cadmium, manganese, cobalt, gold, and copper are used to color glass.

How Do Metals Combine with Other Metals?

Metals do not react with other metals to make new compounds. However, they can melt together to form alloys, which are mixtures of two or more metals. The original metals can be separated from the mixture without a chemical reaction.

One reason transition metals conduct heat and electricity well is their valence electrons are not closely attracted to the nucleus of their individual atoms. In a quantity of a pure metal or an alloy, the electrons freely

Common Properties of Ionic Compounds

Characteristics of ionic compounds include the following:
1. Their strong bonds between ions make them hard to break apart.
2. Most are dull, hard, and brittle solids.
3. Many dissolve in water.
4. They conduct electricity well when melted or dissolved (but not when solid).
5. They have high melting points.
6. They form crystals.

Molybdenite is a compound of molybdenum and sulfur. The silvery black mineral is the main ore for molybdenum. Molybdenum is soft enough to cut with a pocketknife, but it makes steel strong enough to make missile and aircraft parts with.

move from one atom to another. They share their electrical charge equally with all the atoms in the substance. Scientists call this a sea of electrons. The process is called metallic bonding.

How We Use Alloys

Today, steel supports most major buildings. It is also used in cars, trucks, airplanes, ships, high-tech computers, and low-tech staples and paper clips.

Transition metals in alloys strengthen and harden them. Magnesium (Mg) hardens steel. So does molybdenum, which makes steel strong enough for missile and aircraft parts. Niobium strengthens alloys used in pipeline construction and the space program. Ruthenium hardens platinum and palladium to help electrical contacts resist wear. It also hardens platinum and palladium for use as jewelry.

Another important use of alloys made with transition elements is to resist corrosion. Corrosion is a process that dissolves or eats away a substance. (Rust results from one type of corrosion.) Ruthenium helps titanium resist corrosion. In fact, adding just one-tenth of 1 percent ruthenium to titanium makes it a hundred times more resistant than titanium alone. The same tungsten used in electric lightbulbs helps other elements resist corrosion. And chromium improves steel's resistance to corrosion by making it "stainless" steel.

Chapter Five
Elements and You

Transition elements and their compounds pop up in your daily life—even in some places you least expect to find them. Here are some examples.

Around the House

The water heater in your home likely has a zinc coating. The water pipes and wiring are made from copper. And you might have brass door handles made from an alloy of copper and zinc. Transmission wires that bring electricity to the house are made from yttrium. The lightbulbs have tungsten filaments.

Pigments that give color to paint come from a variety of transition metal compounds. Compounds of iron and oxygen make paints red, yellow, and brown. Greens and blues come from different compounds of chromium, copper, or cobalt. Titanium oxide from a reaction of titanium and oxygen makes paint white. (It's also used for white ceramic earthenware, porcelain, and tile.)

In the kitchen, you'll find stainless steel used in the sink, pots and pans, appliances, and knives, forks, and spoons. In the garage, chromium makes the trim on bicycles, cars, and motorcycles shine. Golf clubs contain

Copper tubes are used for refrigeration, air conditioning, and nuclear reactors. Copper also has uses in electrical wiring, plumbing pipes, and cookware. When combined with oxygen, it turns green, like the domes on many state capitol buildings.

titanium and tungsten. Tungsten is also used in golf balls and weights for fishing line.

In the bathroom, parts used in the plumbing fixtures are made with nickel. Razor blades contain cobalt. And toothpaste gets its bright white color from a compound of titanium and oxygen.

In the family room, rechargeable batteries use cadmium or nickel alloys. And handheld games and computers contain copper circuits.

The jewelry box in your parents' bedroom might contain gold, silver, nickel, palladium, rhodium, or platinum. In the garden, molybdenum compounds help plants take in nitrogen and cast off sulfur.

Should You Eat Spinach for Iron?

The cartoon character Popeye is known for encouraging kids to eat spinach. And many people think the green, leafy vegetable is a good source of iron. Spinach does contain iron. Unfortunately, it also contains a chemical that limits the amount of iron the body can use. So, while spinach is certainly not bad for you, it's not a good source of dietary iron.

Transition Metals Away from Home

Because of their high melting points, some transition elements are good choices for hot job, like the exterior of spacecraft returning to Earth's atmosphere. Tungsten and platinum are also used in spacecraft, as well as aircraft and missiles. Platinum coats missile nose cones and jet engine fuel nozzles. Titanium is also used in aircraft and bulletproof armor plating.

Some other uses of transition elements include vanadium and hafnium for the manufacture of tools, and hafnium and zirconium for nuclear reactors. Cobalt is used to detect leaks in pipes. Another form of cobalt kills bacteria in food.

Manhole covers are made from cast iron. If you see a building topped with a green dome, such as many state capitols, you are seeing copper that has reacted with oxygen and carbon dioxide in the atmosphere. The green color indicates copper carbonate. A compound of tungsten and carbon (tungsten carbide) makes high-speed cutting tools and drill bits.

Transition metals and alloys are also used for the arts and decoration. Artisans in ancient Egypt found a way to use gold to beautify objects made from less valuable materials. The process called gilding uses thin sheets of gold hammered into foil. Artists brush the gold foil onto the object, which can look like it is made from solid gold. One

Egyptian king Tutankhamen's coffin is an example of gilding. Ancient artists brushed the wooden coffin with thin sheets of gold foil to make it appear to be made of pure gold.

famous gilded object is King "Tut" Tutankhamen's coffin. It was made of wood and covered with gold foil more than three thousand years ago.

At an art gallery, you'll find bronze sculptures.

In a band or orchestra, brass musical instruments like tubas, trumpets, trombones, and French horns are made from an alloy of copper and zinc. Brass is stronger and has a lower melting point than copper. These qualities make it a good choice for casting, stamping, and pressing into the shapes of the instruments.

Fireworks get some of their colors from transition elements and their compounds. Gold sparkles come from iron; silver ones come from titanium or magnesium. Turquoise comes from copper chloride.

Finding the Iron in Your Cereal

Want to see the iron in your fortified cereal? Here's how:
1. Seal one cup of cereal in a plastic sandwich bag.
2. Crush it into powder with a rolling pin.
3. Place the powder in a nonmetal bowl with a magnet.
4. Add one cup of water. Stir with a plastic spoon for ten minutes.
5. Remove the magnet, and wipe it on a paper towel. The black powdery substance is iron metal.

In the Body

In the proper amounts, some transition metal compounds keep you healthy. But too much of some of them can be dangerous. The transition metals the body needs include chromium, copper, iron, molybdenum, manganese, and zinc.

Chromium helps the brain function. It also helps the body get energy from sugar in the blood. Molybdenum helps iron get into the red blood cells that carry oxygen throughout the body. Copper helps keep bones, nerves, blood vessels, and the immune system healthy. Many of the body's enzymes contain molybdenum or manganese.

Manganese stimulates digestion. Manganese is also important in development of the skeleton and maintaining a healthy nervous system. However, heavy exposure causes a psychiatric condition called "manganese madness." Manganese madness occurs in babies who get too much manganese in soy formula or adults exposed to excessive amounts of manganese at work. Symptoms include irritability, fits of laughter for no reason, hallucinations, violent acts, and sleeplessness. Later in the disease, symptoms include depression, the inability to stay awake, and

The human body needs the same iron atoms used in steel. In the body, iron produces hemoglobin, the molecule that transports oxygen in blood. Too little iron causes anemia, which makes a person weak. Too much damages the kidneys.

shaking, trembling, muscle stiffness, and awkward walking.

Zinc promotes cell growth, helps heal wounds, and helps the immune system work. Zinc oxide serves as sunblock to protect the skin from sunburn.

Transition Metals in Foods

The transition element compounds the body needs come from food. To get chromium, make a salad with romaine lettuce, onions, and tomatoes. Copper is found in lima beans, prunes, mushrooms, potatoes, turnip greens, chicken, turkey, clams, crab, lobster, oysters, shrimp, black-eyed peas, almonds, cashews, walnuts, peanuts, and sunflower seeds.

You can get iron in beef, chicken, seafood, breads, and fortified cereals. Pour milk on your cereal, and you'll add molybdenum. It also comes in bread, peas, and peanuts. Tea is an excellent source of manganese. So are whole-grain products, pineapple, and strawberries. Dietary nickel comes from oatmeal, dried beans, nuts, and chocolate.

Good food sources of zinc include beef, lamb, pork, crab, turkey, chicken, lobster, clams, salmon, and egg yolk.

Another transition metal found in food is the compound titanium dioxide, which is used in white food coloring.

The Periodic Table of Elements

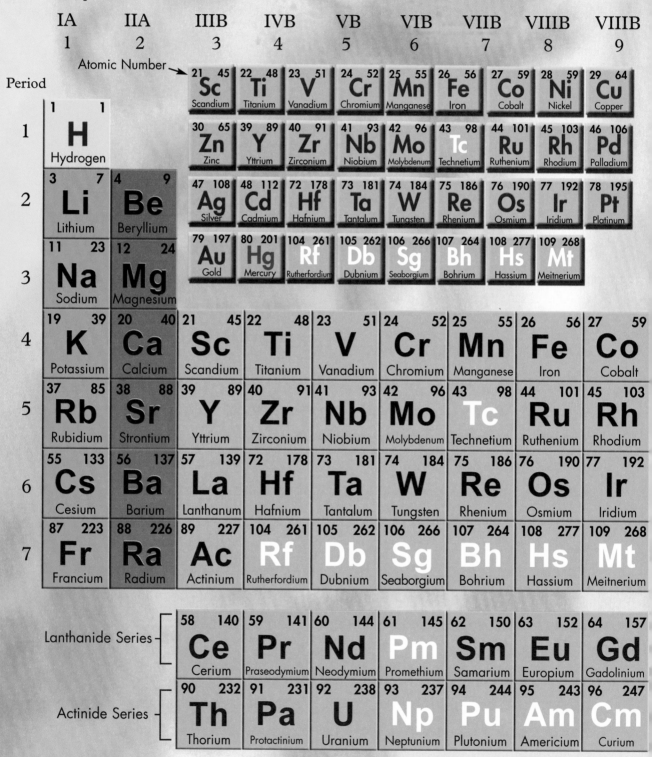

Group

	IA	IIA	IIIB	IVB	VB	VIB	VIIB	VIIIB	VIIIB

Period

Lanthanide Series

58 140	59 141	60 144	61 145	62 150	63 152	64 157
Ce Cerium	Pr Praseodymium	Nd Neodymium	Pm Promethium	Sm Samarium	Eu Europium	Gd Gadolinium

Actinide Series

90 232	91 231	92 238	93 237	94 244	95 243	96 247
Th Thorium	Pa Protactinium	U Uranium	Np Neptunium	Pu Plutonium	Am Americium	Cm Curium

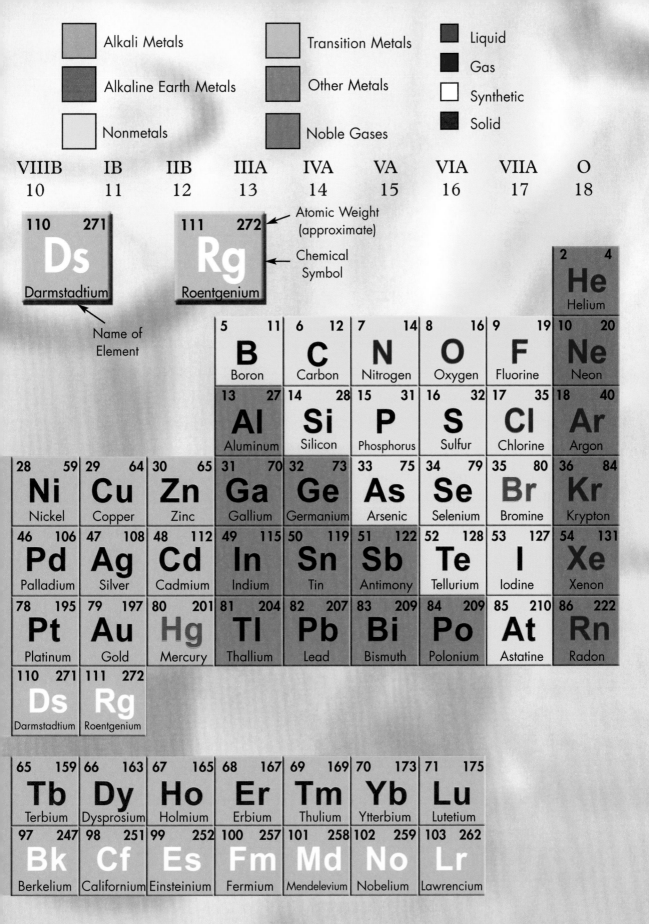

Alkali Metals

Alkaline Earth Metals

Nonmetals

Transition Metals

Other Metals

Noble Gases

Liquid

Gas

Synthetic

Solid

| VIIIB | IB | IIB | IIIA | IVA | VA | VIA | VIIA | O |
| 10 | 11 | 12 | 13 | 14 | 15 | 16 | 17 | 18 |

110	271	111	272
Ds		**Rg**
Darmstadtium		Roentgenium

Atomic Weight (approximate)

Chemical Symbol

Name of Element

								2 4
								He Helium
			5 11	6 12	7 14	8 16	9 19	10 20
			B Boron	**C** Carbon	**N** Nitrogen	**O** Oxygen	**F** Fluorine	**Ne** Neon
			13 27	14 28	15 31	16 32	17 35	18 40
			Al Aluminum	**Si** Silicon	**P** Phosphorus	**S** Sulfur	**Cl** Chlorine	**Ar** Argon
28 59	29 64	30 65	31 70	32 73	33 75	34 79	35 80	36 84
Ni Nickel	**Cu** Copper	**Zn** Zinc	**Ga** Gallium	**Ge** Germanium	**As** Arsenic	**Se** Selenium	**Br** Bromine	**Kr** Krypton
46 106	47 108	48 112	49 115	50 119	51 122	52 128	53 127	54 131
Pd Palladium	**Ag** Silver	**Cd** Cadmium	**In** Indium	**Sn** Tin	**Sb** Antimony	**Te** Tellurium	**I** Iodine	**Xe** Xenon
78 195	79 197	80 201	81 204	82 207	83 209	84 209	85 210	86 222
Pt Platinum	**Au** Gold	**Hg** Mercury	**Tl** Thallium	**Pb** Lead	**Bi** Bismuth	**Po** Polonium	**At** Astatine	**Rn** Radon
110 271	111 272
Ds Darmstadtium	**Rg** Roentgenium

65 159	66 163	67 165	68 167	69 169	70 173	71 175
Tb Terbium	**Dy** Dysprosium	**Ho** Holmium	**Er** Erbium	**Tm** Thulium	**Yb** Ytterbium	**Lu** Lutetium
97 247	98 251	99 252	100 257	101 258	102 259	103 262
Bk Berkelium	**Cf** Californium	**Es** Einsteinium	**Fm** Fermium	**Md** Mendelevium	**No** Nobelium	**Lr** Lawrencium

Glossary

alloy A mixture of two or more metals.

atom The smallest unit of matter that has the characteristics of an element.

atomic number The number of protons in the nucleus of an atom.

atomic weight/atomic mass The sum of the mass of all the atom's subatomic particles.

catalyst A substance that helps start or speed up a chemical reaction.

compound The new substance that results from a chemical reaction.

distillation Taking advantage of different boiling points for different substances, the process heats a mixture until one of the substances in it evaporates, leaving the other as a liquid.

electrolysis The use of electric current to create chemical reactions that separate a molten or dissolved substance that contains both metallic and nonmetallic ions.

electron cloud The space around an atom's nucleus where electrons are likely to be.

froth flotation A refining process where water and air are added to crushed ore that helps unwanted impurities attach to bubbles and rise to the surface, where they can easily be removed.

fusion A process of bringing together the nuclei of two atoms.

ion An atom with unequal numbers of protons and electrons.

refine Reduce a metal in ore to its pure form.

smelting A process that uses heat to separate metal from ore by creating a chemical reaction known as a reduction reaction.

subatomic particles The basic parts of an atom, including neutrons, protons, and electrons.

valence electrons The high-energy electrons that can be lost, gained, or shared during chemical reactions.

For More Information

American Association for the Advancement of Science (AAAS)
1200 New York Avenue NW
Washington, DC 20005
(202) 326-6400
Web site: http://www.aaas.org
This international, nonprofit organization is dedicated to advancing
science and enhancing communication among scientists, engineers,
and the public.

American Chemical Society (ACS)
1155 16th Street NW
Washington, DC 20036
(800) 227-5558
Web site: http://www.acs.org
The ACS is the world's largest scientific society and represents professionals
in all fields of chemistry and sciences that involve chemistry.

American Chemistry Council (ACC)
1300 Wilson Boulevard
Arlington VA 22209
(703) 741-5000
Web site: http://www.americanchemistry.com
The American Chemistry Council represents manufacturers working to protect the environment, public health, and security of the United States.

American Institute of Chemists (AIC)
315 Chestnut Street

Philadelphia, PA 19106

(215) 873-8224

Web site: http://www.theaic.org

This group of professional chemists and chemical engineers offers professional certification and award programs for scientists.

Chemical Heritage Foundation (CHF)

315 Chestnut Street

Philadelphia, PA 19106

(215) 925-2222

Web site: http://www.chemheritage.org

The CHF is dedicated to preserving and promoting the history of chemistry.

Geochemical Society, Department of Earth and Planetary Science, Washington University

One Brookings Drive

St. Louis, MO 63130

(314) 935-4131

Web site: http://www.geochemsoc.org

The Geochemical Society is an international, nonprofit scientific society of faculty, researchers, and graduate students in geochemistry-related fields of study.

Web Sites

Due to the changing nature of Internet links, Rosen Publishing has developed an online list of Web sites related to the subject of this book. This site is updated regularly. Please use this link to access the list:

http://www.rosenlinks.com/uept/tte

For Further Reading

Adler, David. *Cam Jansen and the Mystery of the Gold Coins*. New York, NY: Puffin, 2004.

Avi. *Iron Thunder: The Battle Between the Monitor and the Merrimac*. New York, NY: Hyperion Books for Children, 2007.

DeFelice, Cynthia. *Devil's Bridge*. New York, NY: Avon Books, 2008.

Dingle, Adrian. *The Periodic Table*. Boston, MA: Kingfisher, 2007.

Harding, David, and Moira Johnston, eds. *The Facts On File Chemistry Handbook*. Rev. ed. New York, NY: Facts On File, Inc., 2006.

Hobbs, Will. *Jason's Gold*. New York, NY: Harper Trophy, 2000.

Keller, Rebecca. *Real Science-4-Kids Chemistry Level I*. Albuquerque, NM: Gravitas Publications, Inc., 2005.

Kernaghan, Eileen. *The Alchemist's Daughter*. Saskatoon, Saskatchewan, Canada: Thistledown Press, 2004.

Silvano, Wendi. *Hands-On Chemistry Experiments*. Grand Rapids, MI: McGraw-Hill Children's Publishing, 2004.

Umansky, Kaye. *The Silver Spoon of Solomon Snow*. Cambridge, MA: Candlewick Press, 2005.

Wertheim, Jane, Chris Oxlade, and Corinne Stockley. *The Usborne Illustrated Dictionary of Chemistry*. London, England: Usborne Publishing Ltd., 2006.

Yep, Laurence. *The Journal of Wong Ming-Chung: A Chinese Miner*. New York, NY: Scholastic, 2000.

Zannos, Susan. *Dmitri Mendeleyev and the Periodic Table: Uncharted, Unexplored, and Unexplained*. Hockessin, DE: Mitchell Lane Publishers, 2005.

Bibliography

AboutChemistry.com. Various entries. Retrieved December 11, 2008 (http://chemistry.about.com).

Bauer, Richard C., James P. Birk, and Pamela S. Marks. *A Conceptual Introduction to Chemistry*. New York, NY: McGraw-Hill Higher Education, 2007.

ChemicalElements.com. Various entries. Retrieved June, 24, 2008 (http://www.chemicalelements.com).

Curious About Astronomy. Various entries. Retrieved October 11, 2008 (http://curious.astro.cornell.edu).

Curran, Greg. *Bonding and Molecular Structure*. Franklin Lakes, NJ: Career Press, 2004.

GSI. "How Heavy Can Atomic Nuclei Be?" March 11, 2004. Retrieved October 20, 2008 (http://www.gsi.de/portrait/Broschueren/ Wunderland/03_3.html).

Guch, Ian. *Complete Idiot's Guide to Chemistry*. New York, NY: Alpha Books, 2006.

McGraw-Hill Companies, Inc. *Chemistry Matter and Change*. Columbus, OH: McGraw-Hill Glencoe, 2008.

Moore, John T. *Chemistry for Dummies*. Hoboken, NJ: Wiley Publishing, Inc., 2003.

NASA. "Imagine the Universe." September 5, 2006. Retrieved October 11, 2008 (http://imagine.gsfc.nasa.gov/docs/ask_astro/ answers/961112a.html).

NASA. "Tests of Big Bang: The Light Elements." Retrieved October 11, 2008 (http://map.gsfc.nasa.gov/universe/bb_tests_ele.html).

Wiker, Benjamin D. *The Mystery of the Periodic Table*. Bathgate, ND: Bethlehem Books, 2003.

Index

About the Author

Mary-Lane Kamberg's wedding ring is made from an alloy of palladium and gold called white gold. At the time she took chemistry in high school and college, elements 105–112, 114, 116, and 118 had not yet been discovered. Kamberg is a professional writer and speaker, and she lives in Olathe, Kansas.

Photo Credits

Cover, pp. 1, 17, 31, 40–41 by Tahara Anderson; pp. 5, 23 Andrew Lambert Photography/Photo Researchers; p. 7 © Akio Inoue/amanaimages/Corbis; p. 9 © Borromea/Art Resource, NY; p. 12 © Hulton Archives/Getty Images; p. 14 © Scott Camazine/Photo Researchers; pp. 15, 27 © AP Photos; p. 18 © NASA/JPL-Caltech/Corbis; p. 21 © Time & Life Pictures/Mansell/Time & Life Pictures/Getty Images; p. 25 © www.istockphoto.com/James Steidl; p. 29 © Jacana/Photo Researchers; p. 33 © Joel Arem/Photo Researchers; p. 35 © www.istockphoto.com/Chuck Rausin; p. 37 © Erich Lessing/Art Resource; p. 39 © Jan Bengtsson/Etsa/Corbis.

Designer: Tahara Anderson; Editor: Bethany Bryan;
Photo Researcher: Marty Levick